BRONCHIAL HYGIENE THERAPY:

Modalities& Techniques

BRONCHIAL HYGIENE THERAPY:

Modalities & Techniques

SamehElhabashy
CCRN, MScN, BSc
Critical Care Nursing Department
Faculty of Nursing
Cairo University

PRINCETON
UNIVERSITY
PRESS

2016

BRONCHIAL HYGIENE THERAPY: Modalities & Techniques

First Printing: 2016

ISBN:978-1-365-63212-9

Princeton University Press,

PrincetonUniversity,
41 William Street,
Princeton,
New Jersey 08540
USA webmaster@press.princeton.edu

Special discounts are available on quantity purchases by corporations, associations, educators, and others. For details, contact the publisher at the above listed address.

U.S. trade bookstores and wholesalers: Please contact SamehElhabashy Tel: (+20) 100-7653998; Fax: (+20) 23657190 or email: Sameh17@cu.edu.eg

To my lovely wife and children

Thank you. Without your support and persistence, I would have never accomplished this work.

Content

Acknowledgements

I would like to express my great thanks and appreciation to my family, parents, colleagues, teachers and my students who are always willing to provide their support and guidance. I also appreciate the efforts of (Princeton Press) acquisitions editors and there guidance.

Preface

Pulmonary complications are major causes of morbidity and mortality for patients with compromised airway clearance. Bronchial hygiene therapy referred to any technique utilized to maintain lung sterility by loosening/removing retained secretions for enhancing ventilation and gas exchange. This book focuses on airway clearance techniques and different modalities have been used to assist those who are unable to clear pulmonary secretions effectively, after given a basic background about respiratory system anatomy and physiology. The following information will be useful for respiratory therapist, critical care nurses, and paramedics. I hope that you find this book enjoyable and clinically relevant that because of the simple presentation of ideas and supported figures.

BRONCHIAL HYGIENE THERAPY:
Modalities & Techniques

Introduction:

Bronchial hygiene therapy (formerly referred to as pulmonary toilet) defined as; any technique/procedure utilized to maintain lung sterility by loosening/removing retained secretions, Bronchodilating, and hyper inflating the lung.

Pulmonary complications are major causes of morbidity and mortality for patients with compromised airway clearance. Conditions such as high spinal cord injuries, neuromuscular deficits, or severe fatigue associated with intrinsic lung disease can diminish the effectiveness of a cough, or eliminate the ability to cough altogether. Also, cystic fibrosis, bronchiectasis, and pneumonia can affect the ability of the lungs to manage secretions and influence the viscosity and amount of sputum produced .

Bronchial Hygiene Therapy (BHT) is a broad term used to describe any airway clearance techniques have been used to assist those who are unable to clear pulmonary secretions effectively, including; chest physiotherapy (which consists of postural drainage, percussion, vibration, coughing, and suctioning), breathing exercises, and manual hyperventilation (used in intubated patients), and giving bronchodilators and mucus-thinning medications. The purpose of (BHT) is to improve the clearance of secretions, increase lung volume, maintaining airways, and warm humidity aiming at enhancing ventilation and gas exchange.

1

Glossary:

- Bronchoscopy: direct examination of larynx, trachea, and bronchi using an endoscope
- Cilia: short hairs that provide a constant whipping motion that serves to propel mucus and foreign substances away from the lung toward the larynx
- Crackles: soft, high-pitched, discontinuous popping sounds during inspiration caused by delayed reopening of the airways
- Diffusion: exchange of gas molecules from areas of high concentration to areas of low concentration
- Dyspnea: labored breathing or shortness of breath hemoptysis: expectoration of blood from the respiratory tract
- Hypoxemia: decrease in arterial oxygen tension in the blood
- Hypoxia: decrease in oxygen supply to the tissues and cells
- Orthopnea: inability to breathe easily except in an upright position
- Physiologic dead space: portion of the tracheobronchial tree that does not participate in gas exchange
- Pulmonary perfusion: blood flow through the pulmonary vasculature
- Respiration: gas exchange between atmospheric air and the blood and between the blood and cells of the body
- Ventilation: movement of air in and out of airways
- Wheezes: continuous musical sounds associated with airway narrowing or partial obstruction

The respiratory system is composed of the upper and lower respiratory tracts. Together, the two tracts are responsible for ventilation (movement of air in and out of the airways). The upper tract, known as the upper airway, warms and filters inspired air so that the lower respiratory tract (the lungs) can accomplish gas exchange. Gas exchange involves delivering oxygen to the tissues through the bloodstream and expelling waste gases, such as carbon dioxide, during expiration.

ANATOMY OF THE UPPERRESPIRATORY TRACT

Upper airway structures consist of the nose, sinuses and nasal passages, pharynx, tonsils and adenoids, larynx, and trachea.

Nose

The nose is composed of an external and an internal portion. The external portion protrudes from the face and is supported by the nasal bones and cartilage. The anterior nares (nostrils) are the external openings of the nasal cavities. The internal portion of the nose is a hollow cavity separated into the right and left nasal cavities by a narrow vertical divider, the septum. Each nasal cavity is divided into three passageways by the projection of the turbinates (also called conchae) from the lateral walls. The nasal cavities are lined with highly vascular ciliated mucous membranes called the nasal mucosa. Mucus, secreted continuously by goblet cells, covers the surface of the nasal mucosa and is moved back to the nasopharynx by the action of the cilia (fine hairs). The nose serves as a passageway for air to pass (to and from) the lungs. It filters impurities and humidifies and warms the air as it is inhaled(Fig.1). It is responsible for olfaction (smell) because the olfactory receptors are located in the nasal mucosa. This function diminishes with age.

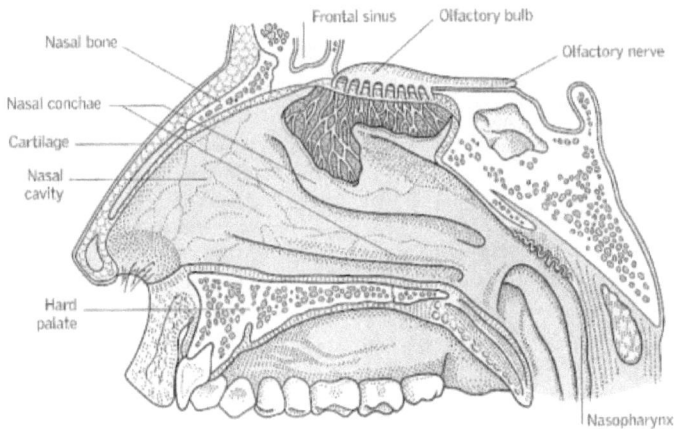

- *Fig. 1: Cross-section of nasal cavity.*

Paranasal Sinuses:

The paranasal sinuses include four pairs of bony cavities that are lined with nasal mucosa and ciliated pseudostratified columnar epithelium. These air spaces are connected by a series of ducts that drain into the nasal cavity. The sinuses are named by their location: frontal, ethmoidal, sphenoidal, and maxillary (Fig.2). A prominent function of the sinuses is to serve as a resonating chamber in speech. The sinuses are a common site of infection. Turbinate Bones (Conchae) The turbinate bones are also called conchae (the name suggested by their shell-like appearance). Because of their curves, these bones increase the mucous

5

membrane surface of the nasal passages and slightly obstruct the air flowing through them.

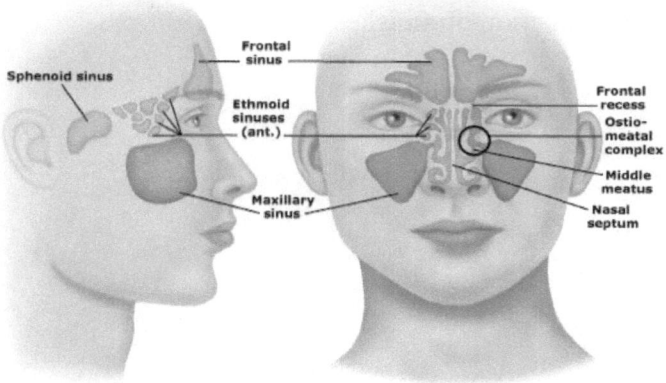

- *Fig. 2: The paranasal sinuses.*

Air entering the nostrils is deflected upward to the roof of the nose, and it follows a circuitous route before it reaches the nasopharynx. It comes into contact with a large surface of moist, warm mucous membrane that catches practically all the dust and organisms in the inhaled air. The air is moistened, warmed to body temperature, and brought into contact with sensitive nerves. Some of these nerves detect odors; others provoke sneezing to expel irritating dust.

Pharynx, Tonsils, and Adenoids:

The pharynx, or throat, is a tubelike structure that connects the nasal and oral cavities to the larynx. It is divided into three regions: nasal, oral, and laryngeal. The nasopharynx is located posterior to the nose and above the soft palate. The oropharynx houses the faucial, or palatine, tonsils (fig. 3). The laryngopharynx extends from the hyoid bone to the cricoid cartilage. The epiglottis forms the entrance of the larynx. The adenoids, or pharyngeal tonsils, are located in the roof of the nasopharynx. The tonsils, the adenoids, and other lymphoid tissue encircle the throat. These structures are important links in the chain of lymph nodes guarding the body from invasion by organisms entering the nose and the throat. The pharynx functions as a passageway for the respiratory and digestive tracts (fig. 3).

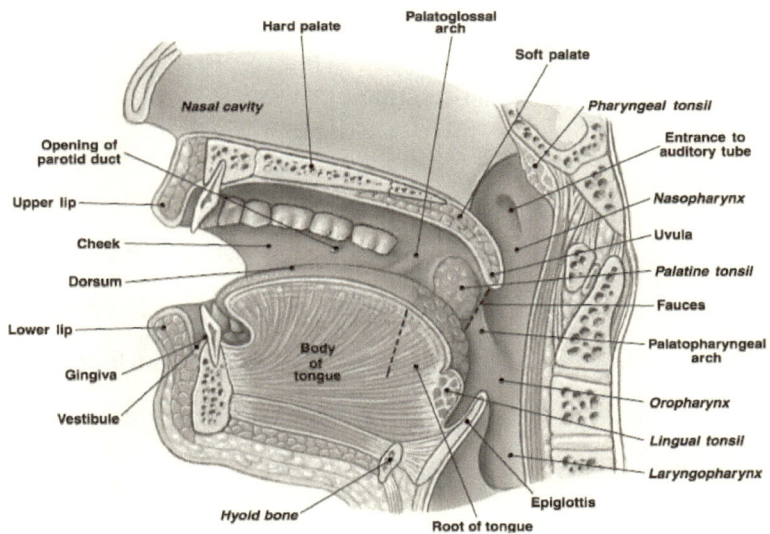

- *Fig. 3: Pharynx, Tonsils, and Adenoids*

Larynx:

The larynx, or voice organ, is a cartilaginous epithelium-lined structure that connects the pharynx and the trachea (fig. 4). The major function of the larynx is vocalization. It also protects the lower airway from foreign substances and facilitates coughing. It is referred to as the voice box and consists of the following:

7

• Epiglottis: a valve flap of cartilage that covers the opening to the larynx during swallowing.

• Glottis: the opening between the vocal cords in the larynx.

• Cricoid cartilage: the only complete cartilaginous ring in the larynx (located below the thyroid cartilage).

• Arytenoid cartilages: used in vocal cord movement with the thyroid cartilage.

• Vocal cords: ligaments controlled by muscular movements that produce sounds; located in the lumen of the larynx.

Trachea:

The trachea, or windpipe, is composed of smooth muscle with C-shaped rings of cartilage at regular intervals. The cartilaginous rings are incomplete on the posterior surface and give firmness to the wall of the trachea, preventing it from collapsing. The trachea serves as the passage between the larynx and the bronchi (fig. 4).

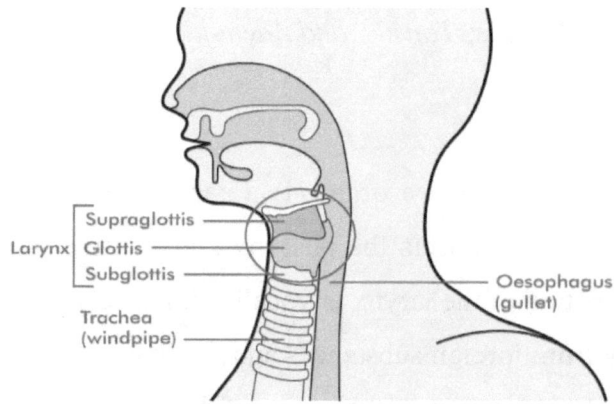

- *Fig. 4: The larynx*

ANATOMY OF THE LOWER RESPIRATORY TRACT:

Lungs:

The lower respiratory tract consists of the lungs, which contain the bronchial and alveolar structures needed for gas exchange.

The lungs are paired elastic structures enclosed in the thoracic cage, which is an airtight chamber with distensible walls (Fig. 5).

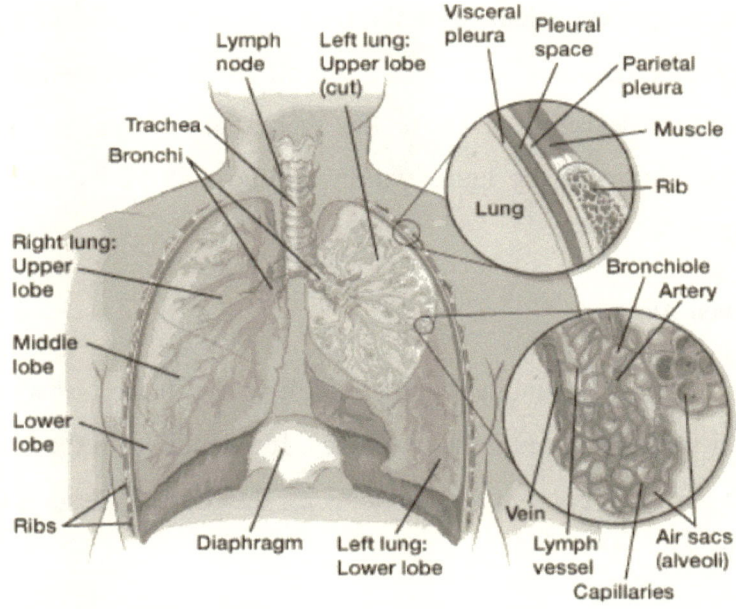

- *Fig. 5: Lower respiratory system.*

Ventilation requires movement of the walls of the thoracic cage and of its floor, the diaphragm. The effect of these movements is alternately to increase and decrease the capacity of the chest. When the capacity of the chest is increased, air enters through the trachea (inspiration) because of the lowered pressure within and inflates the lungs. When the chest wall and diaphragm return to their previous positions (expiration), the lungs recoil and force the air out through the bronchi and trachea. The inspiratory phase of respiration normally requires energy; the expiratory phase is normally passive.

Pleura:

The lungs and wall of the thorax are lined with a serous membrane called the pleura (fig. 5). The visceral pleura cover the lungs; the parietal pleura line the thorax. The visceral and parietal pleura and the small amount of pleural fluid between these two membranes serve to lubricate the thorax and lungs and permit smooth motion of the lungs within the thoracic cavity with each breath.

Mediastinum:

The mediastinum is in the middle of the thorax, between the pleural sacs that contain the two lungs. It extends from the sternum to the vertebral column and contains all the thoracic tissue outside the lungs.

Lobes:

Each lung is divided into lobes. The left lung consists of an upper and lower lobe, whereas the right lung has an upper, middle, and lower lobe (Fig. 6). Each lobe is further subdivided into two to five segments separated by fissures, which are extensions of the pleura.

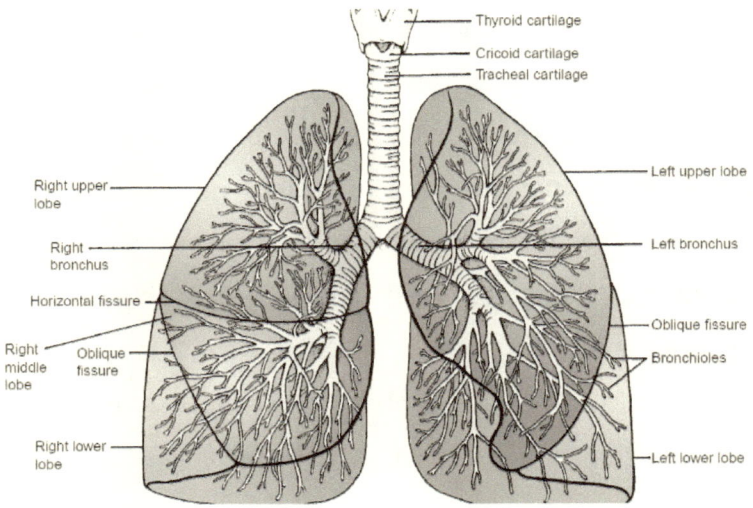

- *Fig. 6: Lobes of the lungs.*

Bronchi and Bronchioles:

There are several divisions of the bronchi within each lobe of the lung. First are the lobar bronchi (three in the right lung and two in the left lung). Lobar bronchi divide into segmental bronchi (10 on the right and 8 on the left), which are the structures identified when choosing the most effective postural drainage position for a given patient. Segmental bronchi then divide into subsegmental bronchi (fig. 5). These bronchi are surrounded by connective tissue that contains arteries, lymphatics, and nerves. The subsegmental bronchi then branch into bronchioles, which have no cartilage in their walls. Their patency depends entirely on the elastic recoil of the surrounding smooth muscle and on the alveolar pressure. The bronchioles contain submucosal glands, which produce mucus that covers the inside lining of the airways. The bronchi and bronchioles are lined also with cells that have

11

surfaces covered with cilia. These cilia create a constant whipping motion that propels mucus and foreign substances away from the lung toward the larynx. The bronchioles then branch into terminal bronchioles, which do not have mucous glands or cilia. Terminal bronchioles then become respiratory bronchioles, which are considered to be the transitional passageways between the conducting airways and the gas exchange airways. Up to this point, the conducting airways contain about 150 mL of air in the tracheobronchial tree that does not participate in gas exchange. This is known as physiologic dead space. The respiratory bronchioles then lead into alveolar ducts and alveolar sacs and then alveoli. Oxygen and carbon dioxide exchange takes place in the alveoli.

Alveoli:

The lung is made up of about 300 million alveoli, which are arranged in clusters of 15 to 20 (fig. 5). These alveoli are so numerous that if their surfaces were united to form one sheet, it would cover 70 square meters—the size of a tennis court. There are three types of alveolar cells. Type I alveolar cells are epithelial cells that form the alveolar walls. Type II alveolar cells are metabolically active. These cells secrete surfactant, a phospholipid that lines the inner surface and prevents alveolar collapse. Type III alveolar cell macrophages are large phagocytic cells that ingest foreign matter (eg, mucus, bacteria) and act as an important defense mechanism.

FUNCTION OF THE RESPIRATORY SYSTEM

The cells of the body derive the energy they need from the oxidation of carbohydrates, fats, and proteins. As with any type of combustion, this process requires oxygen. Certain vital tissues, such as those of the brain and the heart, cannot survive for long without a continuing supply of oxygen. However, as a result of oxidation in the body tissues, carbon dioxide is produced and must be removed from the cells to prevent the buildup of acid waste products. The respiratory system performs this function by facilitating life-sustaining processes such as oxygen transport, respiration and ventilation, and gas exchange.

Oxygen Transport

Oxygen is supplied to, and carbon dioxide is removed from, cells by way of the circulating blood. Cells are in close contact with capillaries, whose thin walls permit easy passage or exchange of oxygen and carbon dioxide. Oxygen diffuses from the capillary through the capillary wall to the interstitial fluid. At this point, it diffuses through the membrane of tissue cells, where it is used by mitochondria for cellular respiration. The movement of carbon dioxide occurs by diffusion in the opposite direction—from cell to blood.

Respiration

After these tissue capillary exchanges, blood enters the systemic veins (where it is called venous blood) and travels to the pulmonary circulation. The oxygen concentration in blood within the capillaries of the lungs is lower than in the lungs' air sacs (alveoli). Because of this concentration gradient, oxygen diffuses from the alveoli to the blood. Carbon dioxide, which has a higher concentration in the blood than in the alveoli, diffuses from the blood into the alveoli. This whole process of gas exchange between the atmospheric air and the blood and between the blood and cells of the body is called respiration (fig. 7).

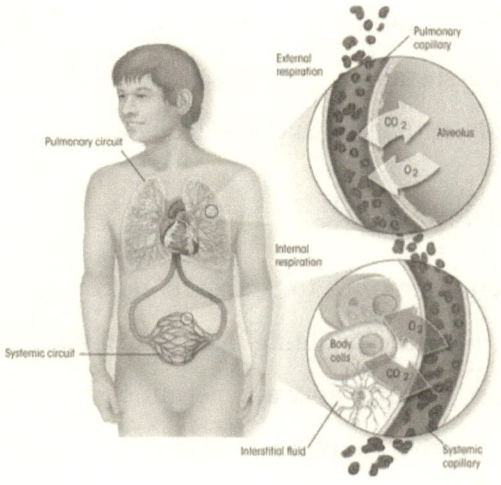

- *Fig. 7: Process of respiration.*

Ventilation

During inspiration, air flows from the environment into the trachea, bronchi, bronchioles, and alveoli. During expiration, alveolar gas travels the same route in reverse (fig.8). Physical factors that govern air flow in and out of the lungs are collectively referred to as the mechanics of ventilation and include air pressure variances, resistance to air flow, and lung compliance.

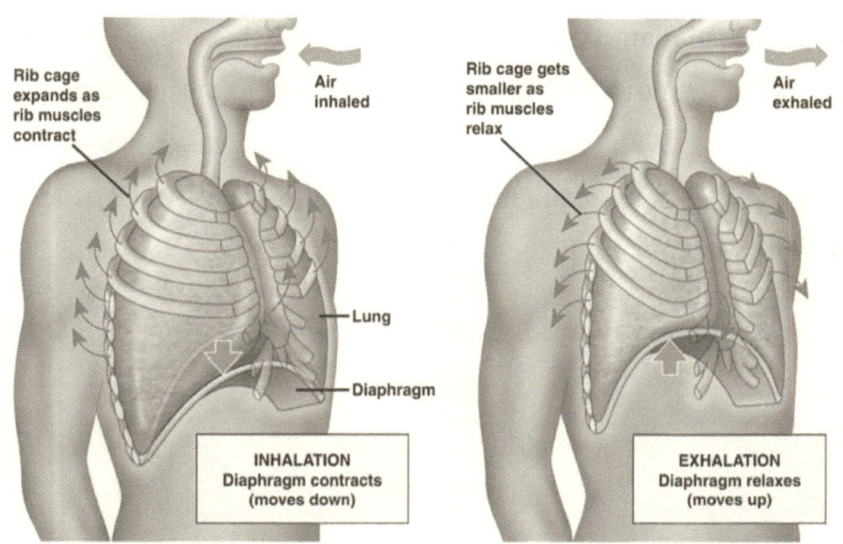

Rib cage expands as rib muscles contract

Air inhaled

Lung

Diaphragm

INHALATION
Diaphragm contracts
(moves down)

Rib cage gets smaller as rib muscles relax

Air exhaled

EXHALATION
Diaphragm relaxes
(moves up)

- *Fig. 8: Process of ventilation (inspiration and expiration).*

AIR PRESSURE VARIANCES

Air flows from a region of higher pressure to a region of lower pressure. During inspiration, movement of the diaphragm and other muscles of respiration enlarge the thoracic cavity and thereby lower the pressure inside the thorax to a level below that of atmospheric pressure. As a result, air is drawn through the trachea and bronchi into the alveoli. During normal expiration, the diaphragm relaxes and the lungs recoil, resulting in a decrease in the size of the thoracic cavity. The alveolar pressure then exceeds atmospheric pressure, and air flows from the lungs into the atmosphere.

AIRWAY RESISTANCE

Resistance is determined chiefly by the radius or size of the airway through which the air is flowing. Any process that changes the bronchial diameter or width affects airway resistance and alters the rate of air flow for a given pressure gradient during respiration (table 1). With increased resistance, greater-than-normal respiratory effort is required by the patient to achieve normal levels of ventilation.

Table 1: Causes of Increased Airway Resistance

Causes of Increased Airway Resistance
Common phenomena that may alter bronchial diameter, which affects airway resistance, include:
• Contraction of bronchial smooth muscle—as in asthma
• Thickening of bronchial mucosa—as in chronic bronchitis
• Obstruction of the airway—by mucus, a tumor, or a foreign body
• Loss of lung elasticity—as in emphysema, which is characterized by connective tissue encircling the airways, thereby keeping them open during both inspiration and expiration

COMPLIANCE

The pressure gradient between the thoracic cavity and the atmosphere causes air to flow in and out of the lungs. When pressure changes are applied in the normal lung, there is a proportional change in the lung volume. A measure of the elasticity, expandability, and distensibility of the lungs and thoracic structures is called compliance. Factors that determine lung compliance are the surface tension of the alveoli (normally low with the presence of surfactant) and the connective tissue (ie, collagen and elastin) of the lungs.

Compliance is determined by examining the volume–pressure relationship in the lungs and the thorax. In normal compliance (1.0 L/cm H2O), the lungs and thorax easily stretch and distend when pressure is applied. High or increased compliance occurs when the lungs have lost their elasticity and the thorax is overdistended (ie, in emphysema). When the lungs and thorax are "stiff," there is low or decreased compliance. Conditions associated with this include pneumothorax, hemothorax, pleural effusion, pulmonary edema, atelectasis, pulmonary fibrosis, and acute respiratory distress syndrome (ARDS), all of which are discussed in later chapters in this unit. Measurement of compliance is one method used to assess the progression and improvement in ARDS. Lungs with decreased compliance require greater-than-normal energy expenditure to achieve normal levels of ventilation. Compliance is usually measured under static conditions.

Lung Volumes and Capacities

Lung function, which reflects the mechanics of ventilation, is viewed in terms of lung volumes and lung capacities. Lung volumes are categorized as tidal volume, inspiratory reserve volume, expiratory reserve volume, and residual volume. Lung capacity is evaluated in terms of vital capacity, inspiratory capacity, functional residual capacity, and total lung capacity. These terms are described in Table 2.

Table 2: Lung Volumes and Lung Capacities

TERM	SYMBOL	DESCRIPTION	NORMAL VALUE*	SIGNIFICANCE
Lung Volumes				
Tidal volume	V_T or TV	The volume of air inhaled and exhaled with each breath	500 mL or 5–10 mL/kg	The tidal volume may not vary, even with severe disease.
Inspiratory reserve volume	IRV	The maximum volume of air that can be inhaled after a normal inhalation	3,000 mL	
Expiratory reserve volume	ERV	The maximum volume of air that can be exhaled forcibly after a normal exhalation	1,100 mL	Expiratory reserve volume is decreased with restrictive conditions, such as obesity, ascites, pregnancy.
Residual volume	RV	The volume of air remaining in the lungs after a maximum exhalation	1,200 mL	Residual volume may be increased with obstructive disease.
Lung Capacities				
Vital capacity	VC	The maximum volume of air exhaled from the point of maximum inspiration VC = TV + IRV + ERV	4,600 mL	A decrease in vital capacity may be found in neuromuscular disease, generalized fatigue, atelectasis, pulmonary edema, and COPD.
Inspiratory capacity	IC	The maximum volume of air inhaled after normal expiration IC = TV + IRV	3,500 mL	A decrease in inspiratory capacity may indicate restrictive disease.
Functional residual capacity	FRC	The volume of air remaining in the lungs after a normal expiration FRV = ERV + RV	2,300 mL	Functional residual capacity may be increased with COPD and decreased in ARDS.
Total lung capacity	TLC	The volume of air in the lungs after a maximum inspiration TLC = TV + IRV + ERV + RV	5,800 mL	Total lung capacity may be decreased with restrictive disease (atelectasis, pneumonia) and increased in COPD.

Diffusion and Perfusion

Diffusion is the process by which oxygen and carbon dioxide are exchanged at the air–blood interface. The alveolar–capillary membrane is ideal for diffusion because of its large surface area and thin membrane. In the normal healthy adult, oxygen and carbon dioxide travel across the alveolar–capillary membrane without difficulty as a result of differences in gas concentrations in the alveoli and capillaries.

Pulmonary perfusion is the actual blood flow through the pulmonary circulation. The blood is pumped into the lungs by the right ventricle through the pulmonary artery. The pulmonary artery divides into the right and left branches to supply both lungs. These two branches divide further to supply all parts of each lung.

Normally about 2% of the blood pumped by the right ventricle does not perfuse the alveolar capillaries. This shunted blood drains into the left side of the heart without participating in alveolar gas exchange.

The pulmonary circulation is considered a low-pressure system because the systolic blood pressure in the pulmonary artery is 20 to 30 mm Hg and the diastolic pressure is 5 to 15 mm Hg. Because of these low pressures, the pulmonary vasculature normally can vary its capacity to accommodate the blood flow it receives. When a person is in an upright position, however, the pulmonary artery pressure is not great enough to supply blood to the apex of the lung against the force of gravity. Thus, when a person is upright, the lung may be considered to be divided into three sections: an upper part with poor blood supply, a lower part with maximal blood supply, and a section in between the two with an intermediate supply of blood. When a person lying down turns to one side, more blood passes to the dependent lung. Perfusion also is influenced by alveolar pressure.

The pulmonary capillaries are sandwiched between adjacent alveoli. If the alveolar pressure is sufficiently high, the capillaries will be squeezed. Depending on the pressure, some capillaries completely collapse, whereas others narrow. Pulmonary artery pressure, gravity, and alveolar pressure determine the patterns of perfusion. In lung disease these factors vary, and the perfusion of the lung may become very abnormal.

Ventilation and Perfusion Balance and Imbalance

Ventilation is the flow of gas in and out of the lungs, and perfusion is the filling of the pulmonary capillaries with blood. Adequate gas exchange depends on an adequate ventilation–perfusion ratio. In different areas of the lung, the ratio varies. Alterations in perfusion may occur with a change in the pulmonary artery pressure, alveolar pressure, and gravity. Airway blockages, local changes in compliance, and gravity may alter ventilation.

A ventilation–perfusion V/Q imbalance occurs from inadequate ventilation, inadequate perfusion, or both. There are fourpossible V/Q states in the lung: normal V/Q ratio, low V/Q ratio (shunt), high V˙/Q˙ ratio (dead space), and absence of ventilation and perfusion (silent unit) (figure 9).Ventilation and perfusion imbalance causes shunting of blood, resulting in hypoxia (low cellular oxygen level). Shunting appears to be the main cause of hypoxia after thoracic or abdominal surgery and most types of respiratory failure. Severe hypoxia results when the amount of shunting exceeds 20%. Supplemental oxygen may eliminate hypoxia, depending on the type of V/Q imbalance.

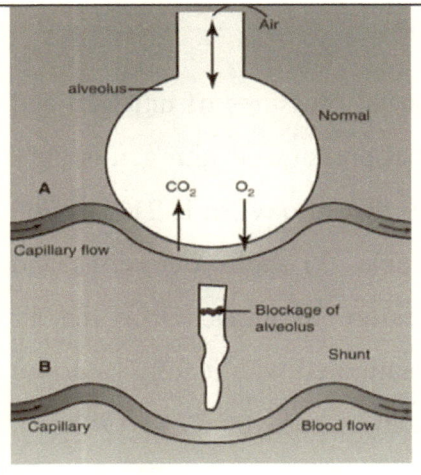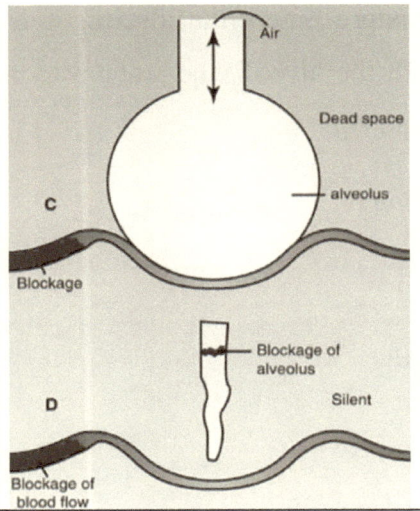	**Normal Ratio (A)**
	In the healthy lung, a given amount of blood passes an alveolus and is matched with an equal amount of gas (**A**). The ratio is 1:1 (ventilation matches perfusion).
	Low Ventilation–Perfusion Ratio: Shunts (B)
	Low ventilation–perfusion states may be called shunt-producing disorders. When perfusion exceeds ventilation, a shunt exists (**B**). Blood bypasses the alveoli without gas exchange occurring. This is seen with obstruction of the distal airways, such as with pneumonia, atelectasis, tumor, or a mucus plug.
	High Ventilation–Perfusion Ratio: Dead Space (C)
	When ventilation exceeds perfusion, dead space results (C). The alveoli do not have an adequate blood supply for gas exchange to occur. This is characteristic of a variety of disorders, including pulmonary emboli, pulmonary infarction, and cardiogenic shock.
	Silent Unit (D)
	In the absence of ventilation and perfusion or with limited ventilation and perfusion, a condition known as a silent unit occurs (D). This is seen with pneumothorax and severe acute respiratory distress syndrome.

- *Fig. 9: ventilation–perfusion V/Q imbalance*

Gas Exchange:

The air we breathe is a gaseous mixture consisting mainly of nitrogen (78.62%) and oxygen (20.84%), with traces of carbon dioxide (0.04%), water vapor (0.05%), helium, and argon. The atmospheric pressure at sea level is about 760 mm Hg. Partial pressure is the exerted by each type of gas in a mixture of gases.

PARTIAL PRESSURE OF GASES:

Based on these facts, the partial pressures of nitrogen and oxygen can be calculated. The partial pressure of nitrogen is 79% of 760 (0.79 × 760), or 600 mm Hg; that of oxygen is 21% of 760 (0.21 × 760), or 160 mm Hg. (Table. 3) spells out terms and abbreviations related to partial pressure of gases. Once the air enters the trachea, it becomes fully saturated with it fully saturates a mixture of gases at the body temperature of 37°C (98.6°F). Nitrogen and oxygen are responsible for the remaining 713 mm Hg (760 − 47) pressure. Once this mixture enters the alveoli, it is further diluted by carbon dioxide. In the alveoli, the water vapor continues to exert a pressure of 47 mm Hg.

Table 3: Partial Pressure Abbreviations

Partial Pressure Abbreviations
P = pressure
PO2 = partial pressure of oxygen
PCO2 = partial pressure of carbon dioxide
PAO2 = partial pressure of alveolar oxygen
PACO2 = partial pressure of alveolar carbon dioxide
PaO2 = partial pressure of arterial oxygen
PaCO2 = partial pressure of arterial carbon dioxide
Pv–O2 = partial pressure of venous oxygen
Pv–CO2 = partial pressure of venous carbon dioxide
P50 = partial pressure of oxygen when the hemoglobin is 50% saturated

PARTIAL PRESSURE IN GAS EXCHANGE

When a gas is exposed to a liquid, the gas dissolves in the liquid until an equilibrium is reached. The dissolved gas also exerts a partial pressure. At equilibrium, the partial pressure of the gas in the liquid is the same as the partial pressure of the gas in the gaseous mixture. Oxygenation of venous blood in the lung illustrates this point. In the lung, venous blood and alveolar oxygen are separated by a very thin alveolar membrane.

Oxygen diffuses across this membrane to dissolve in the blood until the partial pressure of oxygen in the blood is the same as that in the alveoli (104 mm Hg). However, because carbon dioxide is a byproduct of oxidation in the cells, venous blood contains carbon dioxide at a higher partial pressure than that in the alveolar. In the lung, carbon dioxide diffuses out of venous blood into the alveolar gas. At equilibrium, the partial pressure of carbon dioxide in the blood and in alveolar gas is the same (40 mm Hg).

EFFECTS OF PRESSURE ON OXYGEN TRANSPORT

Oxygen and carbon dioxide are transported simultaneously dissolved in blood or combined with some of the elements of blood. Oxygen is carried in the blood in two forms: first as physically dissolved oxygen in the plasma, and second in combination with the hemoglobin of the red blood cells. Each 100 mL of normal arterial blood carries 0.3 mL of oxygen physically dissolved in the plasma and 20 mL of oxygen in combination with hemoglobin. Large amounts of oxygen can be transported in the blood because it combines easily with hemoglobin to form oxyhemoglobin: $O_2 + Hgb \leftrightarrow HgbO_2$

The volume of oxygen physically dissolved in the plasma varies directly with the partial pressure of oxygen in the arteries (PaO2). The higher the PaO2, the greater the amount of oxygen dissolved. For example, at a PaO2 of 10 mm Hg, 0.03 mL of oxygen is dissolved in 100 mL of plasma. At 20 mm Hg, twice this amount is dissolved in plasma, and at 100 mm Hg, 10 times this amount is dissolved. Therefore, the amount of dissolved oxygen is directly proportional to the partial pressure, regardless of how high the oxygen pressure rises.

The amount of oxygen that combines with hemoglobin also depends on PaO2, but only up to a PaO2 of about 150 mm Hg. When the PaO2 is 150 mm Hg, hemoglobin is 100% saturated and will not combine with any additional oxygen. When hemoglobin is 100% saturated, 1 g of hemoglobin will combine with 1.34 mL of oxygen. Therefore, in a person with 14 g/dL of hemoglobin, each 100 mL of blood will contain about 19 mL of oxygen associated with hemoglobin. If the PaO2 is less than 150 mm Hg, the percentage of hemoglobin saturated with oxygen is lower. For example, at a PaO2 of 100 mm Hg (normal value), saturation is 97%; at a PaO2 of 40 mm Hg, saturation is 70%.

OXYHEMOGLOBIN DISSOCIATION CURVE

The oxyhemoglobin dissociation curve (fig. 10) shows the relationship between the partial pressure of oxygen (PaO2) and the percentage of saturation of oxygen (SaO2). The percentage of saturation can be affected by the following factors: carbon dioxide, hydrogen ion concentration, temperature, and 2,3-diphosphoglycerate. A rise in these factors shifts the curve to the right so that more oxygen is then released to the tissues at the same PaO2. A reduction in these factors causes the curve to shift to the left, making the bond between oxygen and hemoglobin stronger, so that less oxygen is given up to the tissues at the same PaO2. The unusual shape of the oxyhemoglobin dissociation curve is a distinct advantage to the patient for two reasons:

1. If the arterial PO2 decreases from 100 to 80 mm Hg as a result of lung disease or heart disease, the hemoglobin of the arterial blood remains almost maximally saturated (94%) and the tissues will not suffer from hypoxia.

2. When the arterial blood passes into tissue capillaries and is exposed to the tissue tension of oxygen (about 40 mm Hg), hemoglobin gives up large quantities of oxygen for use by the tissues.

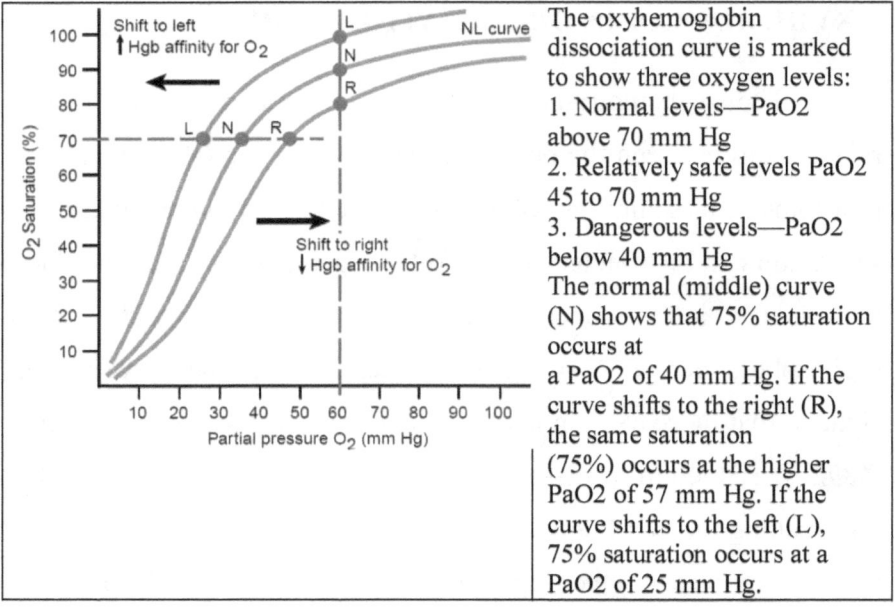

The oxyhemoglobin dissociation curve is marked to show three oxygen levels:
1. Normal levels—PaO2 above 70 mm Hg
2. Relatively safe levels PaO2 45 to 70 mm Hg
3. Dangerous levels—PaO2 below 40 mm Hg
The normal (middle) curve (N) shows that 75% saturation occurs at a PaO2 of 40 mm Hg. If the curve shifts to the right (R), the same saturation (75%) occurs at the higher PaO2 of 57 mm Hg. If the curve shifts to the left (L), 75% saturation occurs at a PaO2 of 25 mm Hg.

-   *Fig. 10: ventilation–perfusion V/Q imbalance*

Clinical Significance:

The normal value of PaO2 is 80 to 100 mm Hg (95% to 98% saturation). With this level of oxygenation, there is a 15% margin of excess oxygen available to the tissues. With a normal hemoglobin level of 15 mg/dL and a PaO2 level of 40 mm Hg (oxygen saturation 75%), there is adequate oxygen available for the tissues but no reserve for physiologic stresses that increase tissue oxygen demand. When a serious incident occurs (eg, bronchospasm, aspiration, hypotension, or cardiac dysrhythmias) that reduces the intake of oxygen from the lungs, tissue hypoxia will result.

An important consideration in the transport of oxygen is cardiac output, which determines the amount of oxygen delivered to the body and which affects lung and tissue perfusion. If the cardiac output is normal (5 L/min), the amount of oxygen delivered to the body per minute is normal. If cardiac output falls, the amount of oxygen delivered to the tissues also falls. In fact, only 250 mL of oxygen is used per minute. Under normal conditions, this is approximately (25%) of available oxygen. The rest of the oxygen returns to the right side of the heart, and the PaO2 of venous blood drops from 80 to 100 mm Hg to about 40 mm Hg.

Carbon Dioxide Transport:

At the same time that oxygen diffuses from the blood into the tissues, carbon dioxide diffuses in the opposite direction (ie, from tissue cells to blood) and is transported to the lungs for excretion. The amount of carbon dioxide in transit is one of the major determinants of the acid–base balance of the body. Normally, only (6%) of the venous carbon dioxide is removed, and enough remains in the arterial blood to exert a pressure of 40 mm Hg. Most of the carbon dioxide (90%) enters the red blood cells; the small portion (5%) that remains dissolved in the plasma (PCO2) is the critical factor that determines carbon dioxide movement in or out of the blood.

In summary, the many processes involved in respiratory gas transport do not occur in intermittent stages; rather, they are rapid, simultaneous, and continuous.

Neurologic Control of Ventilation

Resting respiration is the result of cyclical excitation of the respiratory muscles by the phrenic nerve. The rhythm of breathing is controlled by respiratory centers in the brain. The inspiratory and expiratory centers in the medulla oblongata and pons control the rate and depth of ventilation to meet the body's metabolic demands.

The apneustic center in the lower pons stimulates the inspiratory medullary center to promote deep, prolonged inspirations. The pneumotaxic center in the upper pons is thought to control the pattern of respirations. Several groups of receptor sites assist in the brain's control of respiratory function. The central chemoreceptors are located in the medulla and respond to chemical changes in the cerebrospinal fluid, which result from chemical changes in the blood. These receptors respond to an increase or decrease in the pH and convey a message to the lungs to change the depth and then the rate of ventilation to correct the imbalance.

The peripheral chemoreceptors are located in the aortic arch and the carotid arteries and respond first to changes in PaO2, then to PaCO2 and pH. The Hering–Breuer reflex is activated by stretch receptors in the alveoli. When the lungs are distended, inspiration is inhibited; as a result, the lungs do not become overdistended. Also, proprioceptors in the muscles respond to body movements, such as exercise, causing an increase in ventilation. Thus, range-of-motion exercises in an immobile patient stimulate breathing.

Baroreceptors, also located in the aortic and carotid bodies, respond to an increase or decrease in arterial blood pressure and cause reflex hypoventilation or hyperventilation. carbon dioxide. At approximately age 50, the alveoli begin to lose elasticity. A decrease in vital capacity occurs with loss of chest wall mobility, thus restricting the tidal flow of air. The amount of respiratory dead space increases with age. These changes result in a decreased diffusion capacity for oxygen with age, producing lower oxygen levels in the arterial circulation.

Elderly people have a decreased ability to move air rapidly in and out of the lungs. Gerontologic changes in the respiratory system are summarized in (Table 4). Despite these changes, in the absence of chronic pulmonary disease, elderly people are able to carry out activities of daily living, but they may have decreased tolerance for and require additional rest after prolonged or vigorous activity.

Table 4: Age-Related Changes of the Respiratory System

	STRUCTURAL CHANGES	FUNCTIONAL CHANGES	HISTORY AND PHYSICAL FINDINGS
Defense mechanisms (respiratory and nonrespiratory)	↓ Number of cilia and ↓ mucus ↓ Cough and gag reflex Loss of surface area of the capillary membrane Lack of a uniform or consistent ventilation and/or blood flow	↓ Protection against foreign particles ↓ Protection against aspiration ↓ Antibody response to antigens ↓ Response to hypoxia and hypercapnia (chemoreceptors)	↓ Cough reflex and mucus ↑ Infection rate History of respiratory infections, COPD, pneumonia. Risk factors: smoking, environmental exposure, TB exposure
Lung	↓ Size of airway ↑ Diameter of alveolar ducts ↑ Collagen of alveolar walls ↑ Thickness of alveolar membranes ↓ Elasticity of alveolar sacs	↑ Airway resistance ↑ Pulmonary compliance ↓ Expiratory flow rate ↓ Oxygen diffusion capacity ↑ Dead space Premature closure of airways ↑ Air trapping ↓ Expiratory flow rates Ventilation–perfusion mismatch ↓ Exercise capacity ↑ Anteroposterior (AP) diameter	Unchanged total lung capacity (TLC) ↑ Residual volume (RV) ↓ Inspiratory reserve volume (IRV) ↓ Expiratory reserve volume (ERV) ↓ Forced vital capacity (FVC) and vital capacity (VC) ↑ Functional residual capacity (FRC) ↓ PaO_2 ↑ CO_2
Chest wall and muscles	Calcification of intercostal cartilages Arthritis of costovertebral joints ↓ Continuity of diaphragm Osteoporotic changes ↓ Muscle mass Muscle atrophy	↑ Rigidity and stiffness of thoracic cage ↓ Respiratory muscle strength ↑ Work of breathing ↓ Capacity for exercise ↓ Peripheral chemosensitivity ↑ Risk for inspiratory muscle fatigue	Kyphosis, barrel chest Skeletal changes ↑ AP diameter Shortness of breath ↑ Abdominal and diaphragmatic breathing ↓ Maximum expiratory flow rates

BRONCHIAL HYGIENE THERAPY

Bronchial hygiene therapy (BHT), also known as pulmonary toilet, is helpful in preventing and treating pulmonary complications. The primary phases of lung function that BHT aims to improve are ventilation and diffusion (Fig. 11).

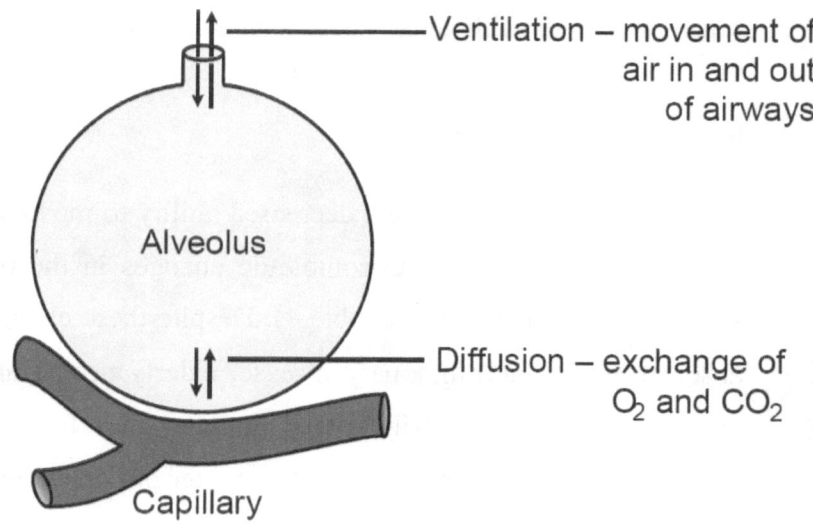

Ventilation – movement of air in and out of airways

Alveolus

Diffusion – exchange of O_2 and CO_2

Capillary

- *Fig. 11: ventilation and diffusion*

These are accomplished through the therapeutic goals: (1) secretion mobilization and removal and (2) improved gas exchange. Specific BHT depends on existing pulmonary dysfunction. The normal airway has a functioning mucociliary "escalator" with a cough reflex and normal mucous production.

The hospitalized patient may have pneumonia, atelectasis, or inability to perform deep breathing, cough, or clear mucus effectively because of weakness, sedation, or pain. The patient may have chronic conditions such as chronic obstructive pulmonary disease (COPD), cystic fibrosis, pulmonary fibrosis, or quadriplegia. The need for and the effectiveness of various methods of BHT are based on physical assessment, chest radiography, measurement of arterial blood gases (ABGs), and additional sources of information as indicated. Any one or a combination of the following measures is used: coughing and deep-breathing maneuvers, airway clearance adjunct devices, chest physiotherapy (CPT), and bronchodilator aerosol therapy.

There are a lot of conditions associated with compromised airway clearance, including congenital disorders (cerebral palsy), respiratory disorders (asthma and chronic bronchitis), postoperative complications (pneumonia), degenerative neuromuscular diseases (muscular dystrophy and amyotrophic lateral sclerosis (ALS), and the use of artificial airways and mechanical ventilation.

COUGHING AND DEEP BREATHING

Effective coughing is necessary for the patient to clear secretions. The objectives of deep breathing and coughing are to promote lung expansion, mobilize secretions, and prevent the side effects of retained secretions (eg, atelectasis and pneumonia). These techniques are effective only if the patient is able to cooperate and has the strength to cough productively.

The patient is positioned seated and upright on the edge of the bed or chair with the feet supported. The nurse instruct the patient to take a slow, deep breath; hold it for 2 to 3 seconds; and exhale slowly for auscultation. If adventitious sounds are auscultated, indicating the presence of secretions, the patient must be made to maximally inhale and cough. Even if secretions are not auscultated, the patient should be encouraged to cough and deep breathe as a prophylactic measure every hour. The patient must be taught the effective use of the incentive spirometer (IS) to have immediate visual feedback on the breath depth, and coached to increase the volume.

Ideally, the patient uses the IS hourly while awake, completing 10 breaths each session followed by coughing, and then the patient progressively increases breath volumes. The nurse coaches the patient to maximize the deep breaths, followed by coughing, and documents the IS volume results. IS, along with coughing and deep-breathing exercises, improves inhaled volumes and prevents atelectasis.

Benefits of breathing deeply:

- Moves air down to the bottom areas of the lungs.
- Opens air passages and moves mucous out (coughing is also easier).
- Helps the blood and oxygen supply to your lungs, boosting circulation.
- Lowers the risk of lung complications such as pneumonia and infections.
- Coughing helps bring up mucous from deep within your lungs.

Techniques of coughing and deep breathing:

1. Get yourself into a comfortable position such as: lying on your back with your knees bent, lying on your side or sitting up in a seated position.

2. Place your hands on your stomach. Take a deep breath in through your nose. Continue until your lungs feel full of air and you notice your stomach pushing against your hand (fig. 12).

3. Through pursed lips, slowly blow air out in one long, slow breath. When you breathe out, concentrate on making your stomach sink in. Repeat steps one, two and three to **complete five breathing cycles**.

4. Take another deep breath – hold for three seconds then huff out three times. (Huffing is a short sharp pant – imagine that you are trying to create mist on a pane of glass.) On the third huff, cough deeply from the lungs, not the throat.

5. Repeat steps two and four to **complete five coughing exercises**.

6. Until you are walking, these exercises should be done every hour while awake. Ask for pain medication if you are sore and not able to do your coughing exercises.

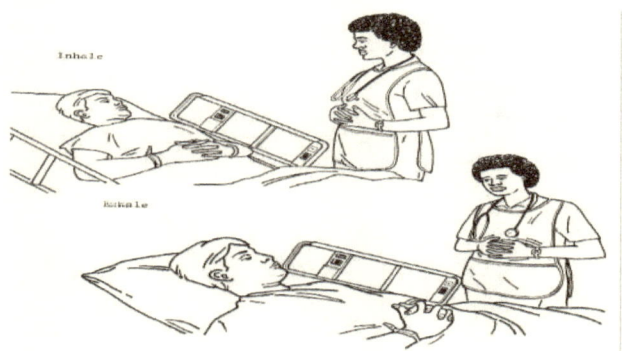

- *Fig. 12: Coughing techniques.*

AIRWAY CLEARANCE

Adjunct Therapies:

Adjunct air way clearance technique used for patients who require mucous removal and, in particular, when coughing efforts are limited by a disease process, injury, or surgery. The Acapella and Flutter valves (fig. 13), two such methods of airway clearance, provide intermittent positive expiratory pressure (PEP) therapy, which improves mucous removal by causing airway vibration to loosen secretions that can be cleared with a cough.

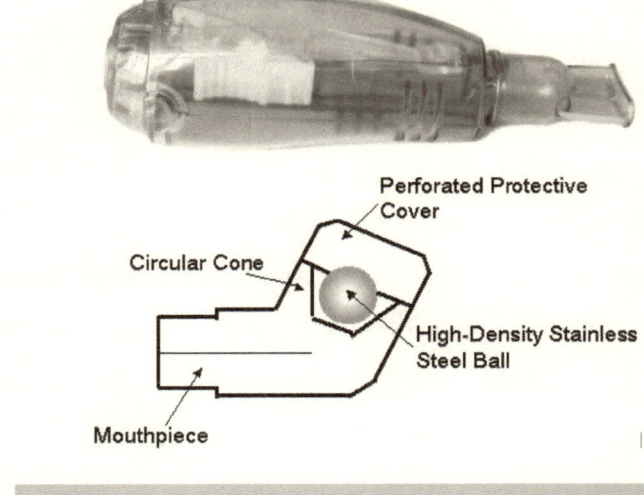

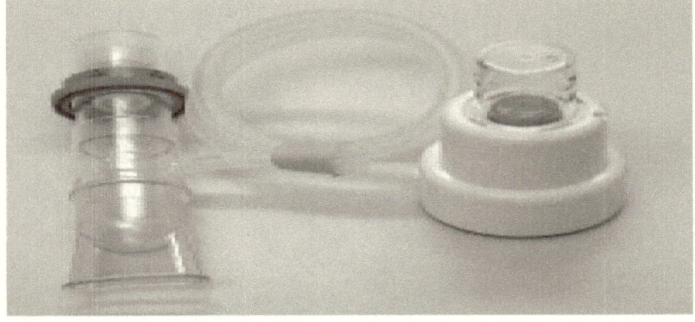

- *Fig. 13:* Acapella and Flutter valves

The Acapella valve is just as effective as the Flutter valve and may be easier to use, especially in elderly patients. Both produce PEP and oscillatory vibrations in the airways to loosen mucus, but the Acapella valve is adjustable for the patient, with two types for patients who can sustain flow of greater than or equal to 15 L/min and less than or equal to 15 L/min; this provides flexibility, especially in those with very low expiratory flows. The nurse assists the patient's cough with positive pressure on the abdominal costal margin duringexhalation to increase the cough force, thus producing a manually assisted cough.

Various specialized BHTs are used for patients with cystic fibrosis and other chronic pulmonary diseases, including autogenic drainage (AD), which may be used with the huff cough. The nurse teaches AD to patients who have reactive airway disease with likelihood of wheezing with normal cough. AD is a series of controlled breaths and uses low-pressure cough with minicoughs instead of one to two big coughs.

Autogenic drainage is a respiratory self-drainage technique that utilizes controlled expiratory airflow (tidal breathing) to mobilize secretions. It consists of three phases (fig. 14):

1. Loosening peripheral secretions by breathing at low lung volumes (slow, deep air movement).
2. Collecting secretions from central airways by breathing at low to middle lung volumes (slow, mid-range air movement).

3. Expelling secretions from the central airways by breathing at mid to high lung volumes (shallow air movements).

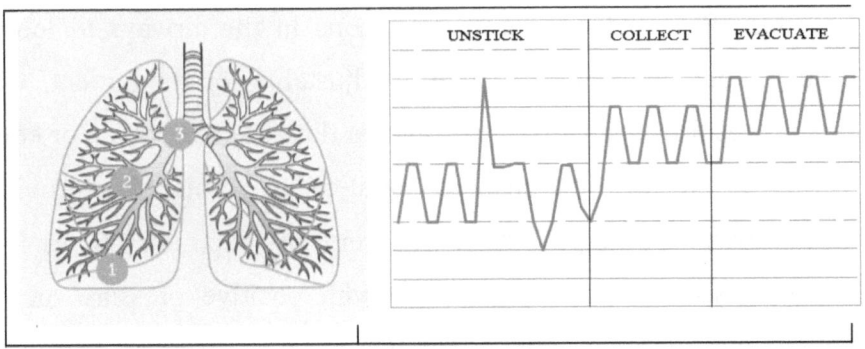

- *Fig. 14: Phases of Autogenic drainage*

The velocity or force of the expiratory airflow must be adjusted at each level of inspiration so that the highest possible airflow is reached in that generation of bronchi, without being high enough to cause the airways to collapse during coughing. Autogenic drainage does not utilize Postural Drainage positions but is performed while sitting upright.

Technique of Autogenic drainage:

1. Choose a breath-stimulating position like sitting or reclining. Relax, with the neck slightly extended.
2. Clear your nose and throat by blowing your nose and huffing.
3. Slowly breathe in through the nose to keep the upper airways open. Use the diaphragm and/or the abdomen if possible.
4. First take a large breath in, hold it for a moment. Breathe all the way out for as long as you can. Now you are at low lung

volume. See picture below. The size of breath and level at which you breathe depends on where the mucus is located.

5. Take a small to normal breath in, and pause. Hold your breath for about 3 seconds. All the upper airways should be kept open. This improves the even filling of all lung parts. The pause allows time for the air to get behind the mucus.

6. Breathe out through the mouth. Keep the upper airways open. This is your glottis, throat and mouth. Breathing out is done in a sighing manner. When you force your breath out the airways can collapse. You will hear a wheeze.

7. At low lung level breathing use your abdominal muscles. Squeeze all the air out until you can breathe out no more.

8. You hear the mucus rattling in the airways when breathing the right way. Put a hand on your upper chest, and feel the mucus vibrating. High frequencies mean that the mucus is in the small airways. Low frequencies mean that the mucus is in the large airways. Using this feedback lets you easily adjust the technique.

9. Repeat the cycle. Inhale slowly to avoid sending the mucus back down. Keep breathing at the low level until the mucus collects and moves upward. Signs of this are:
   o Crackling of the mucus can be heard as you exhale.
   o You feel the mucus moving up.
   o You feel a strong urge to cough.

10. The level of breathing is raised when any of the above occurs. Refer to the picture below. Moving the breathing from lower to higher lung area takes the mucus with it.

11. Finally the collected mucus reaches the large airways where it can be cleared by a high lung volume huff. Don't cough until the mucus is in the larger airways. Cough only if a huff did not move the mucus to the mouth.

12. You have now finished one cycle. Take a break of one to two minutes. Relax and perform breathing control before you start on the next cycle. The cycles are repeated during the session. A session lasts between twenty to forty-five minutes or until you feel all the mucus has been cleared. Do sessions of AD more often if you still have mucus present at the end of a session.

Vest System

The Vest system (Hill-Rom, Batesville, IN), another method of airway clearance, is a chest wall oscillation device that creates chest wall motion through a machine that rapidly alternates air into sections of a vest that is placed circumferentially around the chest (fig. 15). The method is called high- frequency chest wall oscillation, which results in improved secretion removal. This is an alternative to traditional CPT.

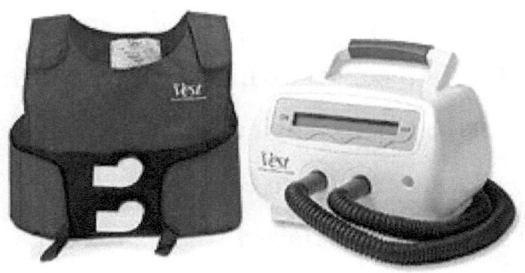

- *Fig. 15: Vest system (Hill-Rom, Batesville, IN).*

The Vest system has been used in trials involving patients with bronchiectasis, cystic fibrosis, COPD, lung transplantation, and even spinal cord injury or quadriplegia, as well as in postoperative intensive care units (ICUs). The Vest system has been shown to improve the removal of mucus and improve pulmonary function. It is well tolerated by surgical patients and can be self-administered at home.

Other therapies include EzPAP, bilevel positive airway pressure (BiPAP), and IS, which may be given before any of the BHTs to improve mucous removal. EzPAP and BiPAP are positive airway pressure devices that enable airway recruitment and prevent atelectasis using between 5 and 20 cm H2O with variable flow of oxygen during therapy.

Both work to reduce atelectasis using positive-pressure therapy, often in combination with aerosol pharmacology agents. EzPAP has only one setting of continuous positive airway pressure (CPAP), and BiPAP has both inspiratory high-pressure and expiratory lower pressure levels (fig. 16). Both are used when IS or other therapies are not sufficient to reduce or prevent atelectasis.

- *Fig. 16: EzPAP with manometer and mask.*

CHEST PHYSIOTHERAPY

Postural drainage, positioning, and chest percussion and vibration are methods of CPT used to augment the patient' efforts and to improve pulmonary function. These may be used in sequence in different lung drainage positions and should be preceded by bronchodilator therapy and followed by deep breathing and coughing or other BHT. Changing the patient's position from supine to upright affects gas exchange, and positioning the patient in the lateral position may improve gas exchange, especially in unilateral lung disease. Positioning the patient with the "good" lung down improves oxygenation.2 This improvement occurs because shunting is decreased when the "good" lung is in the dependent position.

Postural Drainage

Postural drainage positions facilitate gravitational drainage of pulmonary secretions into the main bronchi and trachea based on anatomy of the lung segments (Fig. 17). The focus of postural drainage should be on the lobes affected by atelectasis and on increasing mucous removal with suctioning or by cough effort. Postural drainage is not indicated in all positions for all critically ill patients.

The patient's body is positioned so that the trachea is inclined downward and below the affected chest area. Postural drainage is essential in treating bronchiectasis and patients must receive physiotherapy to learn to tip themselves into a position in which the lobe to be drained. It is done for at least three times daily for up to 30 minutes. It can be done on the night to reduce coughing at night (although PD should be avoided after meals) or in the morning to clear accumulated secretions during the night. Bronchodilators can be used 15 minutes before PD is done to maximize its benefits.

The most affected area is drained first to prevent infected secretions spilling into healthy lung. Drainage time varies but each position requires 10 minutes. If an entire hemithorax is involved each lobe has to drained individual but a maximum of 3 position per session is consider sufficient. The procedure is discontinued if the patient complains of headache, discomfort, dizziness, palpitations, fatigue and dyspnea. Patients may be dyspnic after the various manuovers as the head down position increases work of breathing reduces tidal volume and decreases FRC.

The treatment is often used in conjunction with a technique for loosening secretions in the chest cavity called chest percussion. Chest percussion is performed by clapping the back or chest with a cupped hand. Alternatively, a mechanical vibrator may be used in some cases to facilitate loosening of secretions.[4] There are drainage positions for all segments of the lung. These positions can be modified depending on the patient's condition.Postural drainage may be followed by breathing exercises to help expel loosened secretions from the airway.

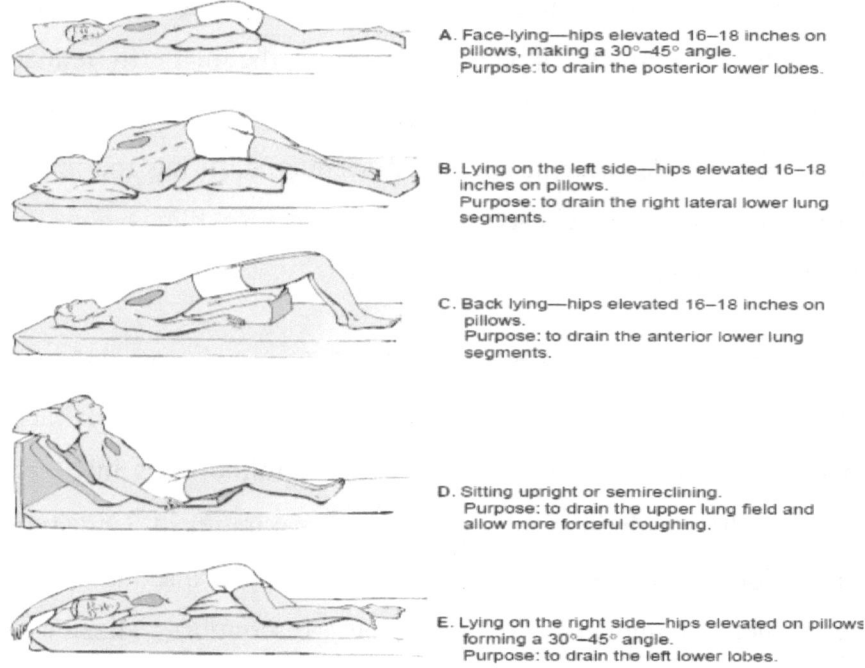

A. Face-lying—hips elevated 16–18 inches on pillows, making a 30°–45° angle.
Purpose: to drain the posterior lower lobes.

B. Lying on the left side—hips elevated 16–18 inches on pillows.
Purpose: to drain the right lateral lower lung segments.

C. Back lying—hips elevated 16–18 inches on pillows.
Purpose: to drain the anterior lower lung segments.

D. Sitting upright or semireclining.
Purpose: to drain the upper lung field and allow more forceful coughing.

E. Lying on the right side—hips elevated on pillows forming a 30°–45° angle.
Purpose: to drain the left lower lobes.

- *Fig. 17: Postural Drainage.*

Contraindications are listed in (table 4). The nurse must closely monitor the patient who is in a head-down position for aspiration, respiratory distress, and dysrhythmias. Alternate techniques include percussion and using a mechanical precursor to stimulate mucous movement while avoiding surgical areas.

Table 4: Contraindications of Postural Drainage.

• Increased intracranial pressure (ICP)
• After meals/during tube feeding
• Inability to cough
• Hypoxia/respiratory instability
• Hemodynamic instability
• Decreased mental status
• Recent eye surgery
• Hiatal hernia

Chest Percussion and Vibration:

Chest percussion (tapotement) and vibration, performed by a trained health care professional, are used to dislodge secretions. Percussion involves striking the chest wall with the hands formed into a cupped shape by flexing the fingers and placing the thumb tightly against the index finger (fig. 18). The patient's position depends on the segment of lung to be percussed. A towel or pillowcase is draped over the area to be percussed, and percussion is performed for 3 to 5 minutes per position. Percussion is never performed over the spine, over the sternum, or below the thoracic cage. Percussion and vibration are performed only on the rib cage. If performed correctly, percussion does not hurt the patient or redden the skin. A clapping sound (as opposed to slapping) indicates correct hand position. Mechanical percussors are also available.

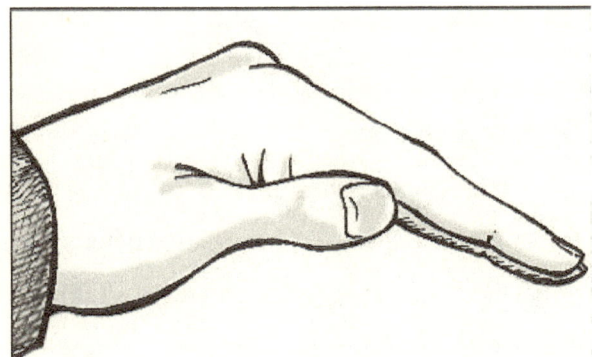

-    *Figure 18: Correct hand shape for percussion*

Vibration takes place during a prolonged pursed-lip exhalation. It increases the velocity and turbulence of exhaled air toloosen secretions. This technique is accomplished by placing the hands side by side with fingers extended and applying the flat of the palm over the affected chest area. The patient inhales deeply and then slowly exhales. While the patient exhales, the nurse vibrates the patient's chest by quickly contracting and relaxing the arm and shoulder muscles.

Vibration is used instead of percussion if the chest wall is extremely painful. Modern ICU beds have options to percuss, vibrate, or provide continuous lateral rotation therapy (CLRT), using either an added module or a system integrated into the bed. These bed features can be used to provide BHT to critically ill patients who may not tolerate manual therapy. The nurse assesses patients for tolerance to both position changes and the level of therapy because most bed systems allow variable settings of high to low frequency of percussion or vibration. Continuous lateral rotation is effective for certain patients, especially ventilated patients.

During vibration, caregivers will place a flat hand firmly against the patient's chest wall, on top of the appropriate lung segment to be drained (fig.19). They then stiffens their arm and shoulder, apply light pressure and create a shaking movement, similar to that of a mechanical vibrating device. If you're the patient, you'll be instructed to breathe in deeply during vibration therapy, and exhale slowly and completely.

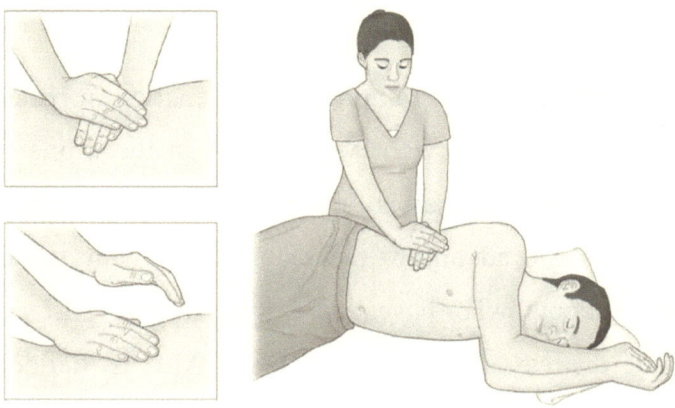

- *Figure 19: Vibration & percussion technique.*

Taking a deep breath and then exhaling slowly and forcefully without straining will help stimulate a productive cough. Also, there is many contraindications of chest percussion/vibration should be avoided (table 5).

Table 5: Contraindications of Chest Percussion and Vibration:

• Fractured ribs/osteoporosis
• Chest/abdominal trauma or surgery
• Bronchopleural fistula
• Pulmonary hemorrhage or embolus
• Chest malignancy/mastectomy
• Pneumothorax/subcutaneous emphysema
• Cervical cord trauma
• Tuberculosis
• Pleural effusions/empyema
• Pulmonary edema
• Asthma
• Fractured ribs/osteoporosis
• Chest/abdominal trauma or surgery

Patient Positioning

Studies demonstrate improved oxygenation in patients with acute respiratory failure who were placed in the prone position, although this maneuver may not ultimately improve survival. Prone positioning is an advanced technique used with critically ill ventilated patients who have acute lung injury (ALI) or acute respiratory distress syndrome (ARDS). ALI is defined by a PaO_2/FiO_2 ratio less than 300. ARDS is defined by a PaO_2/FiO_2 ratio less than 200.

The enhanced oxygenation is attributed to recruitment of collapsed lung areas related to body position change, allowing dependent lung regions to have improved perfusion and ventilation. Prone positioning involves multiple personnel and specialized beds or equipment, and it should be performed only by specially trained staff to prevent the many complications related to prone positioning. Patients who are ventilated benefit from having the head of the bed (HOB) elevated 30 degrees at all times.

The rationale is to promote lung expansion, prevent the aspiration that can occur in the recumbent position in intubated patients, and prevent ventilator-associated pneumonia (VAP). Keeping the HOB elevated 30 degrees, with the associated reduction in VAP, is included in the ventilator bundle to prevent VAP (fig. 20) and is part of the Institute for Healthcare Improvement's 5 Million Lives Campaign. Mobilization of the patient contributes to improved oxygenation, secretion removal, and airway patency.

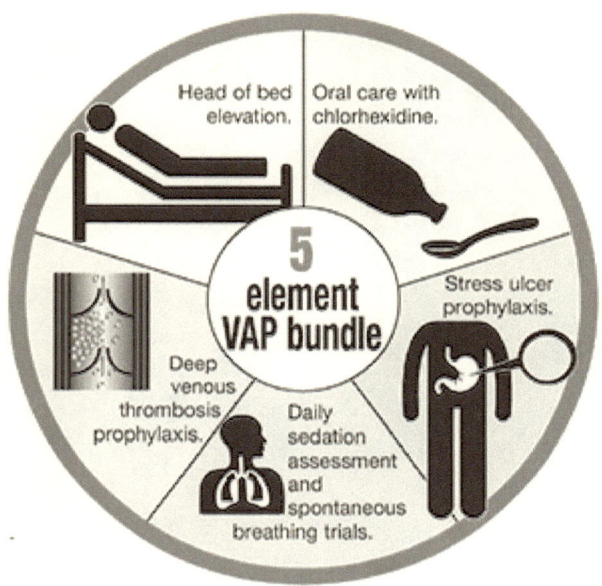

- *Figure 20: VAP Bundle*

Using lateral rotational therapy beds is more effective than the inconsistent nursing care of turning every 2 hours at minimum. (See the American Association of Critical-Care Nurses [AACN] Protocols for Practice, Care of Mechanically Ventilated Patients, 2nd edition, for a complete explanation of prone therapy.) Mobilization of the patient using CLRT improves oxygenation and blood flow to the lung tissue in affected regions (fig.21). CLRT is defined as continuous lateral positioning of less than 40 degrees for 18 of 24 hours daily. The lateral positioning improves blood flow and ventilation in the superior lung regions. CLRT may help reduce incidences of pneumonia, although it may not reduce days on the ventilator or the length of hospital stay.

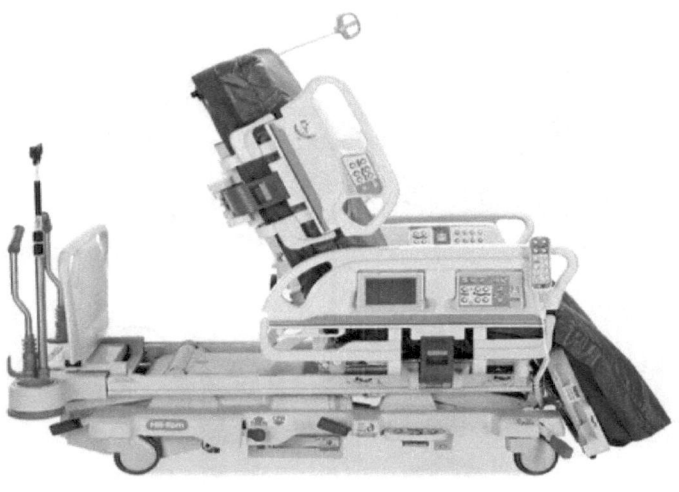

- *Figure 21: continuous lateral rotation therapy (CLRT) bed.*

The results do not show improved survival for ventilated patients using either prone positioning or rotational therapy. However, with CLRT, the improve in the cost associated with VAP, and with prone positioning, there is improved oxygenation. The rotation should be at the maximum to each side, and rotation should be continuous for 18 of 24 hours to obtain best outcomes. Rotation therapy with kinetic therapy refers to beds that rotate greater than or equal to 40 degrees and to CLRT beds that rotate to less than 40 degrees.

Both types of beds may include percussion and vibration modules that allow frequent use of those functions to further improve secretion mobilization. Positioning of critically ill patients remains a nursing intervention for ventilated patients not only to improve oxygenation but also to prevent pressure ulcers. The additional benefit of using CLRT over conventional patient position changes by nursing staff is prevention of skin ulceration. Turning by nurses has become more significant, with the Joint Commission-mandated safety goal of requiring the risk assessment and reassessment for pressure ulcers.

Turning every 2 hours allows the nurse to assess pressure points on the torso and extremities, including the back of the head, and this is even more important in patients with low perfusion. Guidelines should be in place for the use of a scale such as the Braden scale to assess for pressure ulcer risk.The Braden scale (table 6) is a tool used to reassess increased risk factors daily; its use should be followed by consulting the wound care team and providing additional actions to treat pressure ulcers. Repositioning manually, using CLRT or prone positioning, requires care to avoid causing tissue injury when positioning for extended periods.

- Table 6: The Braden scale

1. Completely limited: Unresponsive	2. Very limited: Responds only to painful stimuli.	3. Slightly limited: Responds to verbal commands, cannot ask to turn	4. No impairment: no sensory deficit limiting expression of discomfort
Moisture, *degree to which skin is exposed to moisture*			
1.Constantly moist	2.Very moist	3.Occasionally moist	4.Rarely moist
Activity, *degree of physical activity*			
1. Bedfast	2. Chair fast	3. Walks occasionally	4. Walks often
Mobility, *ability to change and Control Position*			
1. Completely immobile	2. Very limited	3. slightly limited	4. No limitation
Nutrition, *usual food intake pattern*			
1. Very poor	2. Probably inadequate	3. Adequate	4. Excellent
Friction and shear			
1. Problem	2. Potential problem	3. No apparent problem	
15-18 Low Risk, 13-14 Moderate Risk, 10-12 High Risk, >9 Very High Risk			TOTAL

Prolonged positioning in any one position leaves a patient at risk for developing a pressure ulcer, and turning can result in the dislodging of various tubes or lines. The critical care staff should be well trained in prone positioning, monitoring tubes and lines during rotation therapy, and preventing prolonged pressure in lateral positions. Low-pressure airflow mattresses may help reduce skin ulcer occurrence but should not be relied on as a primary prevention method.

Mobilization of ventilated patients using rotation therapy specialty beds is one method for nurses to improve patient outcomes of improved oxygenation; this technique also helps prevent VAP and skin ulcers. Ultimately, the critically ill patient should progress to weight-bearing positions, sitting up in a chair, and, with physical therapy, to ambulation that improves overall physical reconditioning toward a return to independent functioning.

PATIENT SUCTION:

The upper airway warms, cleans and moistens the air we breathe. The trach tube bypasses these mechanisms, so that the air moving through the tube is cooler, dryer and not as clean. In response to these changes, the body produces more mucus. Suctioning clears mucus from the tracheostomy tube and is essential for proper breathing. Also, secretions left in the tube could become contaminated and a chest infection could develop. Avoid suctioning too frequently as this could lead to more secretion buildup.

Suctioning is important to prevent a mucus plug from blocking the tube and stopping the patient's breathing. Suctioning should be considered; any time the patient feels or hears mucus rattling in the tube or airway, In the morning when the patient first wakes up, When there is an increased respiratory rate (working hard to breathe), Before meals, Before going outdoors, Before going to sleep.

The secretions should be white or clear. If they start to change color, (e.g. yellow, brown or green) this may be a sign of infection. If the changed color persists for more than three days or if it is difficult to keep the tracheostomy tube intact, call your surgeon's office. If there is blood in the secretions (it may look more pink than red), you should initially increase humidity and suction more gently.

A Swedish or artificial nose (HME), which is a cap that can be attached to the tracheostomy tube, may help to maintain humidity. The cap contains a filter to prevent particles from entering the airway and maintains the patient's own humidity. Putting the patient in the bathroom with the door closed and shower on will increase the humidity immediately. If the patient coughs up or has bright red blood mucus suctioned, or if the patient develops a fever, call your surgeon's office immediately.

Technique of suction:

Use a clean suction catheter when suctioning the patient. Whenever the suction catheter is to be reused, place the catheter in a container of distilled/sterile water and apply suction for approximately 30 seconds to clear secretions from the inside. Next, rinse the catheter with running water for a few minutes then soak in a solution of one part vinegar and one part distilled/sterile water for 15 minutes. Stir the solution frequently. Rinse the catheters in cool water and air-dry. Allow the catheters to dry in a clear container. Do not reuse catheters if they become stiff or cracked.Connect the catheter to the suction connection tubing.

Lay the patient flat on his/her back with a small towel/blanket rolled under the shoulders. Some patients may prefer a sitting position which can also be tried.Wet the catheter with sterile/distilled water for lubrication and to test the suction machine and circuit.Remove the inner cannula from the tracheostomy tube (if applicable). The patient may not have an inner cannula (fig. 22).

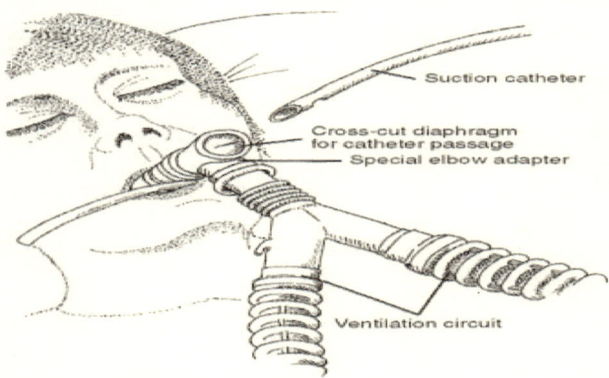

- *Figure 22: Suction procedure.*

There are different types of inner cannulas, so caregivers will need to learn the specific manner to remove their patient's. Usually rotating the inner cannula in a specific direction will remove it. Be careful not to accidentally remove the entire tracheostomy tube while removing the inner cannula. Often by securing one hand on the tracheostomy tubes flange can prevent accidental removal Place the inner cannula in a jar for soaking (if it is disposable, then throw it out).Carefully insert the catheter into the tracheostomy tube. Allow the catheter to follow the natural curvature of the tracheostomy tube. The distance to the location of catheter becomes easier to determine with experience. The least traumatic technique is to pre-measure the length of the tracheostomy tube then introduce the catheter only to that length. For example if the patients tracheostomy tube is 4 cm long, place the catheter 4 cm into the tracheostomy tube.

Often, there will be instances when this technique of suctioning (called tip suctioning) will not clear the patients secretions. For those situations, the catheter may need to be inserted several mm beyond the end of the tracheostomy tube (called deep suctioning). With experience, caregivers will be able to judge the distance to insert the tracheostomy tube without measuring.Place your thumb over the suction vent (side of the catheter) intermittently while you remove the catheter. Do not leave the catheter in the tracheostomy tube for more than 5-10 seconds since the patient will not be able to breathe well with the catheter in place.Allow the patient to recover from the suctioning and to catch his/her breath. Wait for at least 10 seconds.

Suction a small amount of distilled/sterile water with the suction catheter to clear any residual debris/secretions.Insert the inner cannula from extra tracheostomy tube (if applicable).Turn off suction machine and discard catheter (clean according to step 3 if to be reused). And finally, clean inner cannula (if applicable).

SUMMURY:

Respiration is necessary to sustaining life, and the nurse plays an important role in helping the critically ill patient breathe. The nurse must be knowledgeable and skilled in assessing patient needs, providing quick and efficient care, evaluating results of intervention, and supporting and teaching the patient and family. Techniques, equipment, and procedures vary according to the patient's respiratory status. Bronchial hygiene therapy (BHT) is helpful in preventing and treating pulmonary complications.

The goals of CPT are to move bronchialsecretions to the central airways via gravity, external manipulation of thechest, and to eliminate secretions by cough or aspiration with a catheterImproved mobilization of bronchial secretions contribute to improvedventilation-perfusion matching and the normalization of the functional residual capacity and gas exchange. BHT is to improve the clearance of secretions, thereby decreasing airway obstruction and enhancing ventilation and gas exchange. Bronchial hygiene therapy modalities include; Coughing and Deep Breathing, Airway Clearance Adjunct Therapies,Chest Physiotherapy,Patient Positioning, and suctioning.

Notes

References:

-American Nurses Association   code of ethics(O'Neill, 2003).

-Arras, J. (1993). Ethical issues in emergency care.Clinics in geriatric medicine, 9(3), 655-664.

-Billeter, Adrian T.; Druen, Devin; Franklin, Glen A.; Smith, Jason W.; Wrightson, William; Richardson, J. David (2013). "Video-assisted thoracoscopy as an important tool for trauma surgeons: a systematic review". Langenbeck's Archives of Surgery 398 (4): 515–523. doi:10.1007/s00423-012-1016-7. ISSN 1435-2443.

-Brun-Buisson, C., Minelli, C., Bertolini, G., Brazzi, L., Pimentel, J., Lewandowski, K., ...& ALIVE Study Group. (2004). Epidemiology and outcome of acute lung injury in European intensive care units.Intensive care medicine, 30(1), 51-61.

- Calhoon, J. H., &Trinkle, J. K. (1997). Pathophysiology of chest trauma.Chest surgery clinics of North America, 7(2), 199-211.

-Casey, R. &Emde, K. (2008). Displaced fractured sternum following blunt chest trauma.Journal of Emergency Nursing. 34(1), 83-85.

-Clancy, K., Velopulos, C., Bilaniuk, J. W., Collier, B., Crowley, W., Kurek, S., Lui, F., Nayduch, D., Sangosanya, A., Tucker, B. & Haut, E.R. (2012). Screening for blunt cardiac injury: An Eastern Association for the Surgery of Trauma practice management guideline.Journal of Trauma and Acute Care Surgery, 73, S301-S306.

-Collins, J. (2000). Chest wall trauma. Journal of Thoracic Imaging, 15(2),112–119.

-Dalal, S., Nityasha, V. M., &Dahiya, R. S. (2009). Prevalence of chest trauma at an apex institute of North India: A retrospective study. Internet J Surg, 18, 1.

-Demirhan, R., Onan, B., Oz, K., & Halezeroglu, S. (2009). Comprehensive analysis of 4205 patients with chest trauma: a 10-year experience. Interactive cardiovascular and thoracic surgery, 9(3), 450-453.

-Flynn, M. B., &Bonini, S. (1999) Blunt chest trauma: Case report.Critical Care Nurse, 19(5), 68–77.

-Huggins JT, Sahn SA (2004). "Causes and management of pleural fibrosis".Respirology 9 (4): 441–7. doi:10.1111/j.1440-1843.2004.00630.x. PMID 15612954.

-Karlet, M. C. (1997). Update for nurse anesthetists: Thoracic trauma.American Association of Nurse Anesthetists Journal, 65(1), 73–80.

-Kramer, J. L. (2002). Pathophysiology of Thoracic Trauma.In Seminars in Cardiothoracic and Vascular Anesthesia (Vol. 6, No. 2, pp. 57-61).SAGE Publications.

-Kulshrestha, P., Munshi, I., & Wait, R. (2004). Profile of chest trauma in a level I trauma center. Journal of Trauma and Acute Care Surgery, 57(3), 576-581.

-Light RW (2010). "Pleural effusion in pulmonary embolism".SeminRespirCrit Care Med 31 (6): 716–22. doi:10.1055/s-0030-1269832. PMID 21213203.

-LoCicero 3rd, J., & Mattox, K. L. (1989). Epidemiology of chest trauma.The Surgical clinics of North America, 69(1), 15-19.McElroy S.,(2012)

-Pape, H. C., Remmers, D., Rice, J., Ebisch, M., Krettek, C., &Tscherne, H. (2000). Appraisal of early evaluation of blunt chesttrauma: Development of a standardized scoring system for initialclinical decision making. Journal of Trauma-InjuryInfection&Critical Care, 49(3), 496–504.

-Reay, G., & Rankin, J. A. (2013). The application of theory to triage decision-making.International emergency nursing, 21(2), 97-102.

-Rousset, P.; Rousset-Jablonski, C.; Alifano, M.; Mansuet-Lupo, A.; Buy, J.-N.; Revel, M.-P. (2014). "Thoracic endometriosis syndrome: CT and MRI features". Clinical Radiology 69 (3): 323–330. doi:10.1016/j.crad.2013.10.014. ISSN 0009-9260.

-Weldon, Erin; Williams, Jen (2012). "Pleural Disease in the Emergency Department". Emergency Medicine Clinics of North America 30 (2): 475–499.doi:10.1016/j.emc.2011.10.012.

-WHO, (2012); Annual injury surveillance report Egypt, available at:www.emro.who.int/dsaf/dsa1087.pdf

-Wolf, L., Brysiewicz, P., LoBue, N., Heyns, T., Bell, S. A., Coetzee, I., ...&Hangula, R. (2012). Developing a framework for emergency nursing practice in Africa. African Journal of Emergency Medicine, 2(4), 174-181.

-Yahia, A., Ali, N. S., &Elhabashy, S. (2013). Factors Affecting Validity of Arterial Blood Gases Results among Critically Ill Patients: Nursing Perspectives. *Journal of Education and Practice*, *4*(15), 43-56.

-Ragavan AJ, Evrensel CA, Krumpe PK: Interactions of airflow oscillations,tracheal inclination, and mucus elasticity significantly improves simulated cough clearance. Chest 137(2):355–361, 2010

-Pierce LNB: Management of the mechanically ventilated patient, 2nd ed. St. Louis, MO: Saunders, Elsevier, 2007

-Staudinger T, Bojic A, Hozinger U, et al: Continuous lateral rotation therapy to prevent ventilator-associated pneumonia. Crit Care Med 38(2): 486–490, 2010

-Daniels T: Physiotherapeutic management strategies for the treatment of cystic fi brosis in adults. J MulitdiscipHealthc 3:201–212, 2010

-Taccone P, Pesenti A, Latini R, et al: Prone positioning in patients with moderate and severe acute respiratory distress syndrome. JAMA 302(18):1977–1984, 2009

-Al-Tawfi q JA, Abed MS: Decreasing ventilator-associated pneumonia in adult intensive care units using the Institute for Healthcare Improvement Bundle. Am J Infect Control 38(7):552–556, 2010

-Swadener-Culpepper L: Continuous lateral rotation therapy. Crit Care Nurse 3(2):55–57, 2010

-Preventing never events: Pressure ulcers. Joint Commission on Perspectives on Patient Safety (4):5–7, 2009

-Bergstrom N, Braden BJ, Lagtuzza A, et al: The Braden scale for predicting pressure sore risk. Nurs Res 36:205–210, 1987

-Halm MA, Krisko-Hagel K: Instilling normal saline with suctioning: Beneficial technique or potentially harmful sacred cow? Am J Crit Care 17(5):469–472, 2008

-Lawrence DM: Procedure 18, chest tube placement (perform). In Lynn- McHale Wiegand DJ, Carlson KK (eds): AACN Procedure Manual forCritical Care, 6th ed. Philadelphia, PA: Elsevier, 2011

-Lee TA, Schumock GT, Bartle B, et al: Mortality risk in patients receiving drug regimens with theophylline for chronic obstructive pulmonary disease. Pharmacotherapy 2(9):1039–1053, 2009

-Gelinas C, Tousignant-Laflamme Y, Tanguay A, et al: Exploring the validity of the bispectral index, the critical-care pain observation tool and vital signs for the detection of pain in sedated and mechanically ventilated critically ill adults: A pilot study. Intensive Crit Care Nurs 27(1):46–52, 2011

-Checkley W, Brower R, Korpak A, et al: Effects of a clinical trial on mechanical ventilation practices in patients with acute lung injury. Am J RespirCrit Care Med 177(11):1215–1222, 2008

-Esan A, Hess DR, Raoof S, et al: Severe hypoxemic respiratory failure: Part 1 ventilator strategies. Chest 137(5):1203–1216, 2010

-Modrykamien A, Chatburn R, Ashton RW: Airway pressure release ventilation: An alternative mode of mechanical ventilation in acute respiratory distress syndrome. Cleve Clin J Med 78(2):101–110, 2011

-Griffiths MJ, Finney SJ: Small steps in the right direction for ventilatorinducedlung injury: Prevention, prevention, prevention. Crit Care Med 3(1):196–197, 2010

- Sawyer RG, Tache Leon C: Common complications in the surgical intensive care unit. Crit Care Med 38(9):S483–S493, 2010tion of blood.